Crisfield Johnson

The one Great Force

The Cause of Gravitation, Planetary Motion, Heat, Light, Electricity, Magnetism...

Crisfield Johnson

The one Great Force
The Cause of Gravitation, Planetary Motion, Heat, Light, Electricity, Magnetism...

ISBN/EAN: 9783743419636

Manufactured in Europe, USA, Canada, Australia, Japa

Cover: Foto ©berggeist007 / pixelio.de

Manufactured and distributed by brebook publishing software
(www.brebook.com)

Crisfield Johnson

The one Great Force

THE ONE GREAT FORCE:

THE CAUSE OF

GRAVITATION, PLANETARY MOTION, HEAT, LIGHT,

ELECTRICITY, MAGNETISM,

CHEMICAL AFFINITY, AND OTHER NATURAL PHENOMENA.

———— • • • ————

BY CRISFIELD JOHNSON.

———— • • • ————

PROPOSITION:—*The One Great Force of the Material Universe is the Self-Repulsion of Caloric, acting on the Inertia of Ordinary Matter.*

———— • • • ————

BUFFALO:
PUBLISHED BY BREED & LENT.
———
1868.

THOMAS, HOWARD & JOHNSON,
Stereotypers, Printers and Binders.
BUFFALO, N. Y.

PREFACE.

THIS is a large subject and a small book; parts of the former must be insufficiently treated, and, brief as is the latter, yet its few pages may perchance contain many untenable statements, and some foolish ones. Its sole object, however, is to establish the proposition enunciated on the title page, and some mistakes in treating of the minor manifestations of force, (should such be found,) ought not to invalidate the theory, if there be still sufficient evidence of its truth—of which I have no shadow of doubt.

Let no man be barred from investigation by the authority of NEWTON, for that authority is *against* the views of gravitation generally attributed to him, as appears by many of his writings, and especially by his third letter to Bentley, from which I have quoted in Chapter I. At least, read his views before you reject mine.

Believers in an inherent principle of gravitation, acting inversely as the square of the distance, are also referred to the whole of Dr. Faraday's essay on the Conservation of Force, a few extracts from which appear in the body of this work. It has been republished in this country, together with other essays on kindred subjects, including two very important ones by Mr. Grove and Dr. Carpenter, in a work edited by E. S. Youmans, and entitled "The Conservation and Correlation of Forces." Dr. F. there shows at full length, and with the most convincing logic, the unsoundness of the ordinary (not the Newtonian) views of gravitation; and, when gravitation falls, all other kinds of (apparent) attraction fall with it.

Particular attention is called to the explanation, on the following pages, of *Capillary Attraction*, of *Tides*, of *Elasticity*, and of *Muscular Action*.

C. J.

Willink, Erie Co., N. Y., Sept. 3, 1868.

CONTENTS.

THE ONE GREAT FORCE.

THE ONE GREAT FORCE.

CHAPTER I.

ATTRACTION IMPOSSIBLE.

For many years, the belief has been steadily increasing among the leaders of the scientific world that Heat, Light, Electricity, Magnetism and Chemical Affinity are all manifestations of the same force, all convertible into each other, and that the sum of all these forces, like the sum of all the matter in the universe, is neither increased nor diminished by the workings of nature or the acts of man, and it is now the almost universal belief of men of science that Deity alone, by a miraculous interference with the course of nature, can change the total amount of force or matter.

The doctrine of the identity of the different modes of force with each other has been named the "Correlation of Forces." For this grand theory we are indebted almost entirely to the genius of Grove, and his bril-

2

liant essays furnish the most conclusive evidence of the truth of this hypothesis. The theory of the unchangeable amount of the sum total is termed the "Conservation of Force." This latter doctrine, especially, is so firmly imbedded in the scientific mind of the age, that Mr. Herbert Spencer takes it as the very foundation of his new system of philosophy. Regarding the "Persistence of Force," as he terms it, he says: "The sole truth which transcends experience, by underlying it, is thus the persistence of force. This, being the basis of experience, must be the basis of any scientific organization of experiences. To this an ultimate analysis brings us down, and on this a rational synthesis must be built up."

Whether it be correct to assume this doctrine as an axiom or not, few will be able to reject it as a demonstrated fact when supported, as it now is, by the experiments of all modern philosophers. But which one of these correlated forces is the cause of the others, or whether the cause of all is to be sought elsewhere, scientific men do not pretend to say. For some unknown reason, too, they do not generally include gravitation in the list of correlated forces. Grove, with the occasional perversity of genius, gives the opinion that the only connection between gravitation and the other forces is that the motion of a falling body, like any other motion, is converted into heat on being resisted, and adds: "With all due deference, I cannot agree with those who seek a more mysterious connection."

Yet this same attribute of matter, gravitation, is its most wonderful characteristic. How or why an unintelligent substance should go in the direction of another substance, or should have the power to make that other substance come toward it, is something entirely incomprehensible and inconceivable. Though Sir Isaac Newton is usually considered the author of the present system of gravitation, yet, in truth, he only discovered the identity of the force which in some way governs the planets, with that which causes an apple to fall to the earth, and did not pretend to attribute the cause of the phenomena he observed to attraction. On the contrary, in one of his letters he wrote thus: "That gravity should be innate, inherent and essential to matter, so that one body should act upon another at a distance, through a vacuum, without the mediation of anything else, by and through which their force may be conveyed from one to another, is to me so great an absurdity that I believe no man who has, in philosophical matters, a competent faculty of thinking, can ever fall into it. Gravity must be caused by an agent, acting constantly according to certain laws, but whether this agent be material or immaterial, I leave to the consideration of my reader."

And again: "What I call attraction, may be performed by impulse, or by some means unknown to me; I use the word here to signify, in general, any force by which bodies tend toward one another, whatever may be the cause." Mr. Grove, in quoting this last remark, says: "How the phenomena are produced, to which the term attraction is applied, is still a mystery;" and so, indeed, it does seem a strange mystery that this inanimate earth should have the power to make a body entirely unconnected with it, come toward it, or that any inanimate body should, by its own force, move in one direction rather than another. Not only is such a power entirely inconceivable to the mind of man, but it is in direct contradiction to the doctrine of the conservation of force, as was clearly shown by the late Dr. Faraday, who, when living, stood first in the list of England's scientific men.

In one of his essays, after giving his assent to the doctrine of the conservation of force, and stating the received opinion of gravitation as "*a simple attractive force*, exerted between any two, or all, the particles or masses of matter, at every sensible distance, with a force varying inversely as the square of the distance," he says: "This idea of gravity appears to me to ignore entirely the principle of the conservation of force; and, by the terms of its definition, if taken in an absolute sense, '*varying* inversely as the square of the distance,' to be in direct opposition to · it, and it now becomes my duty to point out where this contradiction occurs, and to use it in illustration of the principles of conservation. Assume two particles of matter A and B, in free space, and a force in each, or both, by which they gravitate toward each other, the force being unalterable for an unchanging distance, but varying inversely as the square of the distance, when the latter varies. Then, at the distance of ten, the force may be estimated as one, whilst at the distance of one, that is, one-tenth of the former, the force will be one hundred; and, if we supposed an elastic spring to be introduced between the two, as a measure of the attractive force, the power compressing it will be a hundred times as much in the latter case as in the former.

"But from whence can this enormous increase of power come? If we say that it is the character of this force, and content ourselves with that, as a sufficient answer, then, it appears to me, we admit a creation of power, and that to an enormous amount; yet by a change of

condition so small and simple as to fail in leading the
least instructed to think that it can be a sufficient
cause.   We should admit a result which would equal
the highest act our minds can appreciate of the working
of infinite power upon matter; we should let loose the
highest law in physical science which our faculties
permit us to perceive, namely, the *conservation of force*.
Suppose the two particles, 'A' and 'B,' removed
back to the greater distance of ten; then the force of
attraction would be only a hundredth part of what they
previously possessed; this, according to the statement
that the force varies inversely as the square of the
distance, would double the strangeness of the above
results; it would be an annihilation of force—an effect
equal in its infinity and its consequences with creation,
and only within the power of Him who has created."

Again: "The same line of thought grows up in the
mind, if we consider the mutual gravitating action of
one particle and many.   The particle 'A' will attract
the particle 'B,' at the distance of a mile, with a certain degree of force.   It will attract a particle 'C,' at
the same distance, with a power equal to that with
which it attracts 'B.'   If myriads of like particles be
placed at the given distance of a mile, 'A' will attract
each with equal force, and, if other particles be accu-

mulated around it, within and without the sphere of
two miles in diameter, it will attract them all with a
force varying inversely with the square of the distance.
How are we to conceive of this force growing up in
'A' to a million fold or more, and, if the surrounding
particles be then removed, of its diminution in an equal
degree. Or how are we to look upon the power raised
up in all these other particles by the action of 'A' on
them, or by their action on one another, without admit-
ting, according to the limited definition of gravitation,
the facile generation and annihilation of force?" All
the foregoing remarks are applicable to attractions of
every description; not only of gravitation, but also of
cohesive chemical affinity and magnetism.

After showing, in various ways, the absurdity of the
ordinary definition of gravitation, and its inconsistency
with this fundamental idea of the conservation of force,
Dr. Faraday concludes that "the principle of the conservation of force would lead us to assume, that when
'A' and 'B' attract each other less because of increasing distance, then some other exertion of power, either
within or without them, is proportionally growing up;
and, again, that when this distance is diminished,
as from ten to one, their power of attraction, now
increased a hundred fold, has been produced out of

some other form of power, which has been equivalently reduced."

*　　*　　*　　*　　*　　*　　*　　*　　*

"I would much rather incline to the belief that bodies affecting each other by gravitation, act by lines of force of definite amount, or by an ether pervading all space, than admit that the conservation of force can be dispensed with."

This last idea of Dr. F., though vaguely expressed, is undoubtedly the true basis, not only of gravitation, but of all the correlated forces of nature; and the design of this little work is to show the nature of this "ether pervading all space," and *how* it affects matter so as to produce gravitation, heat, light, etc., etc.

Most of the English scientific writers, in addition to the doctrines of the correlation and conservation of forces, have added another, in no wise connected with them, though generally held by the same men, and that is, that the different manifestations of force, heat, electricity, etc., are, in reality, only "modes of motion," analogous, in many respects, to sound, which analogy furnishes the advocates of this theory with many of their strongest arguments. This doctrine probably owes its origin, so far as heat is concerned, to Count Rumford, the distinguished American-born scientist of the last century, and has been adopted by a large number of the scientific men of the present day, although, as will be shown, it is in direct contravention of the established ideas on the conservation of force.

Prof. Tyndall and Mr. Grove are two of the most eminent of the writers who have advocated this theory, but there is a decided difference in their ideas as to how the different kinds of motion are propagated; the first named gentleman believing that they are transmitted in waves through an impalpable ether which pervades all bodies and fills the space between, while the latter claims that the agents, commonly called imponderables, are molecular affections of ordinary matter, and that the interstellar medium, which all admit must exist, though extremely attenuated, has all the attributes of other matter, "especially weight." But weight, itself, is only a tendency to move in a certain direction, and to leave it out of the number of correlated and conserved forces, is actually to destroy both conservation and correlation. None of the writers on these subjects, undertake to say where this same motion comes from, and Mr. Grove insists that there is probably no such thing as finding the ultimate cause of any thing.

But motion is only a transient attribute of matter. Even if we admit, as Mr. G. claims, that matter is always in motion, to a greater or less extent, yet there is certainly a change in the amount manifested by the same article at different times. Now, motion, not being a permanent attribute of ordinary matter, but being superimposed on matter in different degrees, cannot be the end of all research into the subject of force. We must look to some *permanent* attribute or attributes of matter for all ultimate causes, under Providence. Prof. Tyndall seems desirous of thus founding the origin of physical phenomena in some permanent attribute of matter, and undertakes to account for the light and heat of the sun by supposing that numerous ærolites, revolving around that luminary, are continually falling upon its surface with tremendous velocity, and that this motion is transferred to an all embracing ether in waves; which waves, when they reach the nerves of feeling and sight, give the impression of heat and light.

But what makes these ærolites fall to the sun? Attraction; so that the heat of the sun, and all its thousand manifold effects upon the earth, according to Prof. T.'s theory, are ultimately traceable to attraction; to the power of inanimate matter to make other matter unconnected with it come towards it. So, too, the learned Professor accounts for the heat and light of a candle, by supposing that the atoms of oxygen in the air, are attracted with enormous force toward the atoms of carbon in the melted wax or tallow, and that the clashing together of these atoms causes a vibration in the ether, which has been mentioned above, thus producing the phenomena of heat and light, and, as heat and light are universally admitted to be convertible into electricity, magnetism, etc., it necessarily follows that all forms of force result from attraction. And yet, strangely enough, the same writer is a firm adherent of the doctrine of the "Conservation of Force," which the theory of attraction directly contradicts.

The existence of attraction, as a universal attribute of matter, is inconsistent with the known workings of the universe. It acts inversely as the square of the distance; the nearer the stronger. Now, if two particles or two worlds attract each other at all, that attraction must go on increasing until they come in contact. Then no other substance can attract either of them as strongly as they attract each other unless it is equally near; that is, unless it is in contact with them, and then, of course, it cannot attract them apart, but only adds its power of attraction to that of the mass. So that the inevitable result of installing the single force of attraction in nature, would be that, instead of worlds circling around worlds throughout all space, there would be a constant aggregation of all the matter in the universe. Toward the largest body, wherever that might be, atoms, planets, suns, would be hurled with ever increasing speed, constantly adding to the attraction of the mass, and drawing towards it all matter from all the ends of creation. Even if, as Tyndall and others claim, the motion of these falling bodies should be transferred as a quivering motion to the mass, and that motion should be heat, yet there could be no conductor to carry it off, for the interstellar medium would itself be attracted, and the whole would necessarily be converted into one quivering molten mass.

Attraction, then, cannot be the great force of nature, because,

First. It is inconceivable that an unintelligent body should have the power to make another body come towards it, or should be able of itself to go in one direction rather than another.

Second. It is in direct opposition to the established doctrine of the conservation of force.

Third. It is inconsistent with the known workings of the universe.

CHAPTER II.

But, while attraction by inanimate matter is inconceivable, nay, impossible, the action of a self-repellent substance is plain and simple. If a fluid is self-repellent, but not reducible to ultimate atoms, the inner parts, as it expands, continue to act on the outer, driving them away from the center with a force that always remains the same as to the whole, though the amount exerted at any particular point, constantly decreases as it expands, because it acts on a larger surface. On striking against ordinary, inert matter, it would either carry the latter with it or else be turned in its course, and move straight on in the new direction until again interrupted.

Self-repulsion is the only source of motion of which we can conceive as an attribute of unintelligent matter, and it is also the only one which is in harmony with the doctrine of the conservation of force. Self-repulsion, radiating from the center outward, is in strict accordance with the analogies of nature. The apparent examples of attraction, such as magnetism, gravitation, etc., have always had an appearance of mystery to the human mind, and no one has been able to say, positively, whether they proceed from an interior or an exterior force, but, in all cases of motion which we ourselves cause, and which, therefore, we have a better opportunity to investigate, we find it invariably radiating in every direction from its point of origin. Strike an anvil with a hammer and the sound radiates in every direction. Ignite a match and the heat and light radiate in every direction. Generate electricity and it follows the same law, if equally good conductors be furnished in every direction. Yet, if all matter were self-repellent, it is plain that all would fly in every direction, producing universal chaos. This result does not take place, consequently there must be, aside from ordinary matter, some self-repellent fluid, producing the motion which exists throughout the universe.

As the name caloric is tolerably well known throughout the world, it may as well be continued as the name of an eternal, self-repellent fluid, which, by its different manifestations produces the phenomena not only of heat, light, electricity, magnetism and chemical affinity, but also of gravitation, planetary motion, and many other forces. As to heat and light, the fluid may have different attributes producing these effects, though it is more probable that the difference in sensation arises from the nature of the senses themselves; the rapid motion of the fluid producing the effects we call heat and light. But its indispensable attribute is self-repulsion.

As I have before remarked, the general tendency of the same writers, who have so ably advocated the doctrines of the correlation and conservation of force, is to dispense with caloric as a hypothetical fluid, leaving us without any cause for the phenomena of nature, except the impossible and insufficient one of attraction. Some of the arguments adduced to prove its non-existence will be noticed hereafter, but at present we will only say, that, without its self-repellent force, we cannot account for even one of the phenomena of nature.

Whatever may be the case with ordinary matter, it is certain that caloric cannot be resolved into ultimate atoms. An indivisible atom, with definite length, breadth and thickness, could not be self-repellent as to its own substance, for then it could be again divided; and it could not repel other atoms to a sensible distance, because as soon as it ceased to be in contact with them, even by the millionth part of a hair's-breadth, it could no longer operate on them without an expansion of its own substance, which would destroy its definite length, breadth and thickness. It must, then, be infinitely divisible and self-repellent.

[On the other hand, it is very probable, though scarcely susceptible of proof, that all ordinary, inert matter *is* reducible to ultimate atoms, and that these atoms are cubical. If we fuse any substance thoroughly by means of heat, and then suddenly remove the heat, the substance crystalizes; that is, it is arranged in rectilinear forms, and the presumption is strong that these are the original forms of its ultimate particles. This, however, is decidedly speculative.]

Self-repulsion, as an attribute of all matter, would produce unbounded confusion. There must be a conservative, controlling principle, and this principle is found in inertia, which is an attribute of all ordinary matter, in fact of everything except caloric. With this principle, the ordinary ideas of gravitation are totally inconsistent, although most works on natural philosophy attribute both gravitation and inertia to the same matter. But inertia is the quality of remaining in just the condition in which anything is placed, while gravitation is a tendency to move in a certain direction, and the absurdity of attributing both qualities to the same substance is self-evident.

Says the same distinguished authority heretofore quoted, (Dr. Faraday,) "There is one wonderful quality of matter, *perhaps its only true indication*, namely, inertia; but in relation to the ordinary definition of gravity, it only adds to the difficulty. For if we consider .two particles of matter a certain distance apart, attracting each other under the power of gravity and free to approach, they will approach; and when at half the distance, each will have stored up in it, because of its inertia, a certain amount of mechanical force. This must be due to the force exerted, and, if the conservation principle be true, must have consumed an equivalent proportion of the cause of attraction, and yet, according to the definition of gravity, the attractive force is not diminished thereby, but is increased fourfold, the force growing up within itself the more rapidly the more it is occupied in producing other force. On the other hand, if mechanical force from without be used to separate the particles to twice their distance, this force is not stored up in momentum by inertia, but disappears, and three-fourths of the attractive force at the first distance disappears with it. How can this be? We know not the physical condition or action from which inertia results, but inertia is always a pure case of the conservation of force. It has a strict rela-

3

tion to gravity, as appears by the proportionate amount
of the force which gravity can communicate to the
inert body, but it appears to have the same strict
relation to other forces acting at a distance, as those of
magnetism, or electricity, when they are applied by the
tangential balance as to act independently of the gravitating force. It has the like strict relation to forces
communicated by impact, pull, or in any other way. It
enables a body to take up and conserve a given
amount of force until that force is transferred to other
bodies, or changed into an equivalent of some other
form; that is all we perceive in it; and we cannot find
a more striking instance, amongst natural or possible
phenomena, of the necessity of the conservation of
force as a law of nature, or one more in contrast with
the assumed variable condition of the gravitating force
supposed to reside in the particles of matter."

Only one expression in the above extract needs
remark. The "physical condition or action from which
inertia results" is simply the absence of force. It is a
purely negative quality, and, while unlimited self-repulsion would produce only chaotic motion, universal
inertia would involve all nature in the quietude of
death. But the action of a self-repellent fluid upon
inert matter would produce confined, modified, regulated

motion, such as we see throughout the universe, and would be in strict analogy with that dualism which is the most apparent characteristic of nature.

It is evident that, if a self-repellent substance were exerting its force from every side upon a body of inert matter, and could not penetrate the latter, it would have a tendency to hold the particles of the body together. If the caloric penetrated the body, but, instead of returning from it, carried it along in its own course, then close to either side of that body there would be force coming from the outside, with nothing, or with a less force, to oppose it, the force from the side of the body being cut off by it; and therefore small bodies would be driven toward the larger, giving the appearance of attraction, though there would be no more attraction than when water ascends toward the mouth through a tube from which the air has been removed by suction. It is the pressure from the other side that does the work in every case.

I believe, then, and intend in the following chapters to prove that,

THE ONE GREAT FORCE OF THE MATERIAL UNIVERSE, IS THE SELF-REPULSION OF CALORIC, ACTING ON THE INERTIA OF ORDINARY MATTER.

CHAPTER III.

THE velocity of Mercury is about 110,000 miles an hour; that of Venus about 81,000; of the Earth, 75,000; of Mars, 50,000; of Jupiter, 30,000; of Saturn, 25,000; of Herschell, 15,000. This steady decrease of velocity, shows plainly that the planets have been thrown off from the sun, one after the other, the outermost first, and that their original immense velocity, probably derived immediately from explosion, has been decreased by outside pressure. What is the force which, as fast as the planets have been thrown off from the sun, has stayed their outward course and made them circle around the parent luminary? Not attraction, for that is impossible, and, even if it were possible, attraction *in vacuo* would create no friction, and therefore would not retard their velocity.

Caloric in enormous quantities exists throughout all space; its self-repulsion drives it in every direction, and, on every body existing in open space, a tremendous pressure is exerted on every side by this self-repellent element. Our senses are attuned to the amount of caloric in circulation, and a slight change in that amount gives the idea of intense cold or heat, but above and below the ordinary range of our senses are thousands and thousands of degrees, extending from pure fire to the entire absence of heat.

The amount of force is eternal, but, of course, when caloric radiates from any point, the amount exercised on any given body will be inversely as the square of the distance of that body, because the total force must be divided by the area of the surface of a sphere, the center of which is at the radiating point, and the areas of the surfaces of spheres are as the squares of their radii. When a body receives into its interior more of this fluid than it gives out, then on every side of it will be a space in which other bodies will be *driven*, not attracted, towards its center, because the first named body cuts off the force coming from one side, and leaves the force on the other side free to act. This inward driving force is gravitation, and is the stronger the nearer we approach the absorbing body.

The subtle, self-repellent fluid forces its way through the crust of the earth into its interior, creating millions of tiny apertures, which close behind it, preventing, to a great extent, its egress; the next wave on the outside drives in more caloric, and thus on, always increasing the size of the earth by its action on the inside, and driving bodies toward it by its action on the outside.

The sun is acted on in the same way, and, from all parts of surrounding space, streams of caloric are rushing toward the great luminary, not because they are directly attracted, but because there they can most easily escape the surrounding pressure.

If we imagine a cave, filled with furious winds, in the center of which is an immense ball, with an elastic crust, perforated by innumerable valve-like apertures through which the air can force its way to the interior, but can not, at least in as large quantities, escape, we can see at once that the air would tend toward the ball from every direction, nor need we suppose any special attractive force in the ball itself.

Quiescent air would only fill the ball till it was as dense as on the outside, but wind, arising from an independent force, would, by the action of its waves, continually drive the air through the valves, thus constantly increasing the tension of the imprisoned part, and the self-repulsion of caloric is also an independent force, which produces like results on all the heavenly bodies.

Newton demonstrated that the joint action of a continuous rectilinear impulse, and of a central force which decreased inversely as the square of the distance, would produce elliptical motion, similar to the actual orbits of the planets, and as attraction, if it existed at all, would decrease inversely as the square of the distance, Newton's demonstrations have been considered as proving the fact of attraction, notwithstanding his own express rejection of it. But gravitation, arising from the absorption of caloric into any body, would also drive other bodies inward with a force inversely as the square of the distance.

Acting, as caloric does, in free space, the pressure is the same everywhere, and the tendency toward the valvular pores of the sun is the same at the distance of fifty million of miles as at one million. But, at each of these distances, the total amount is distributed over a hollow sphere, the surface of which is in proportion to the square of its radius, and therefore the force exercised at any particular point is inversely as the square of its distance.

This is the explanation of gravitation in harmony with the conservation of force; this dispenses with the attraction *in vacuo*, which Newton repudiated; this is the precise action of "an ether pervading all space," which Faraday sought.

When the amount of this self-repellent fluid, enclosed in the sun or earth, becomes too great for the crust to contain it, it causes an explosion, either great or small. The small explosions are represented on earth by volcanic eruptions and earthquakes; the great ones of the sun fling off a portion either of its crust or of its interior, more probably the latter, which becomes a planet. Explosion, alone, would send it out in a direct line from the center; the rotary motion of the sun would throw it forward at a tangent to the surface; the composition of these two forces must give it an intermediate direction, but the moment it leaves the sun it is subjected to the action of the streams of caloric rushing toward that body from every direction, producing motion around it in elliptical curves, precisely as Newton demonstrated would result from a centrifugal impulse in a straight line, and a centripetal one which should decrease inversely as the square of the distance.

Meanwhile the continual pressure, which drives it from a straight line into an ellipse, constantly retards its velocity, which decreases all the way from more than 110,000 miles per hour to less than 15,000. After ages have passed away, another planet is flung off, and another, and another, each circling in gradually increasing ellipses around the parent orb. Each is subject to the same pressure of caloric, which tends toward the center of the planet, forces its way through the valve-like pores of the crust, increases the size, and finally, by a grand explosion, flings off a moon, which, being operated on by the same forces, moves in an ellipse around the planet, as the latter does around the sun.

In strict accordance with these views we see the planets steadily, though not regularly, increasing in size from mercury outward, and, though all are not attended by moons, yet, in the case of those that are, the number of satellites increases in the same direction.

Contrary to appearances, there is more caloric going toward the sun than coming from it, more coming toward the earth than goes from it, and more coming from the side opposite the sun than from the sun itself. Nevertheless, the greater directness of the concentrated rays of the sun produces more heat and light than the oblique rays from the outside, the same as the more direct rays of summer have greater effect on both animal and vegetable life than those of winter, though in the latter case, the earth is far nearer to the fountain of heat and light, than in the former.

A very strong confirmation of these views is found in the phenomena of capillary attraction, a direct reversal of the laws of gravitation, by which liquids in extremely small tubes rise instead of falling. An immaterial attraction by the earth, if it existed, could be as easily exerted in one kind of tube as another, and, were gravitation the result of a simple attraction, there could be no such thing as capillary attraction. But, while streams of caloric are rushing inward, sufficient to cause gravitation, a lesser number are pouring outward. The light we see coming from our sister planets, whether caused by reflection or emission, is sufficient evidence on this point. The first named rays, coming from all parts of the celestial sphere, concentrate in the earth; the others act close .to their point of emission, and consequently go directly upward. An exceedingly small tube cuts off the concentrating rays from above, and the direct rays from below enter the bottom of the tube and carry the liquid upward.

The fact that when a capillary tube is composed of a bad conductor of caloric, like glass, it need not be so small as when composed of a good conductor like iron, strengthens this theory and adds another to the numerous proofs that gravitation, capillary attraction, and all the other forces of Nature, are the result of the self-repulsion of caloric.

CHAPTER IV.

THE correlation of these forces is more evident than that of any other two. At the point where either originates, that is, where combustion takes place, both can always be discovered to some extent; the chief difference being that, from the same amount of combustion, light can be perceived at a much greater distance than heat, and that, in passing through most solid substances, light disappears, while heat continues. The common origin of the two has been demonstrated by numerous experiments, and, if the doctrine of the correlation and conservation of forces be correct, that origin can only be caloric, a self-repellent fluid, escaping in every direction from the burning substance, producing the sensation of heat on the nerves of feeling, and that of light on the organs of sight. As the eye is evidently a far more delicate structure than any other part of the human system, it can easily be understood that only a slight amount of caloric would

be necessary to produce the sensation called sight, com-
pared with what would be required to stir the grosser
nerves of feeling.   Thus the beacon, whose caloric is
so far dissipated that its effect in exciting appreciable
warmth is lost at the distance of a few rods, has yet
sufficient strength to set in motion the delicate pulsations of the optic nerve, miles and miles from the point
of combustion.

This would be the natural result of the susceptibility
of our visual apparatus, whether the phenomena of heat
and light were caused by the self-repulsion of caloric
or by the communication of motion to the particles of
gross matter.   But there is another fact, connected with
these two forces, which is incompatible with any other
theory than that of an actual fluid, emitted from the
luminous substance.   When an opaque body intervenes
between the point of combustion and the eye, the
sensation of heat is still excited, though that of light
is not.   Now, if heat and light were only motions of
gross matter affecting the different senses, the same
result would follow, as in the case of the distant beacon,
and the most delicate sense would be the most easily
affected.   When the waves of sound are obstructed by
an obstacle, the person with the finest sense of hearing
will most easily catch the murmur; on the contrary,

when the waves of light and heat are obstructed by an opaque substance, it is the coarser nerves of feeling that are affected by the amount of force which struggles through the obstacle.

The only way we can account for these contrary results is to suppose that there is an actual substance, caloric, liberated from all combustibles, which, in its natural condition, affects the nerves both of sight and feeling, and which instantly, on its liberation, is driven with tremendous rapidity in every direction by its own self-repellent force. Through glass and other transparent substances it passes in small quantities, (which accounts for their being such poor conductors of heat,) but it passes pure, uncontaminated by other matter, and therefore readily affects the infinitesimal nerves of the eye. Transparent substances, in order to transmit, unimpaired, the images of different objects from one side to the other, must be perforated by innumerable tiny apertures, perfectly straight, and so close together that the caloric apparently passes in a continuous stream, though, in fact, between all these millions of passages, there is a solid substance which we can feel and weigh. In the case of fluids and gases, it is probable that the caloric has force enough to drive the particles slightly aside, thus making for itself a straight

road to the eye.   But, in the case of an opaque substance, although a much larger quantity of caloric may
gain the further side, yet it contracts a taint, it carries
with it accompanying particles of the gross matter
through which it has made its way, causing it to be
too coarse to affect the sense of sight, though it readily
makes itself felt by the less delicate nerves of feeling,
and, by its self-repulsive power, lifts the mobile column
of the thermometer.

Thus, when wood burns in a stove, caloric escapes
from it, which is then capable of affecting the senses
as heat and light.   As it passes through the iron it
carries with it an infinitesimal part of the latter, which
prevents its affecting the eyes as light.   It may seem
at first incredible that fire should for years pass
through a piece of iron, all the while carrying a portion with it, without utterly destroying the metal.   In
time its structure *is* destroyed, or "burnt out," but we
have an example in the circulation of odors, how infinitesimal parts of a body may be carried away for a
long time, apparently without injuring its structure or
diminishing its size.   A piece of India-rubber, for instance, will affect the sense of smell for years, and yet
suffer no sensible diminution, though we know that, all
the while, infinitely small particles are being carried

away by the surrounding air; because we know that the sense of smell is only affected by actual contact with the things smelled. So, too, a solution of a grain of certain coloring matter may be diluted and sub-diluted, until as much as will hold the two-billionth part of a .grain in solution may be examined by a microscope, and still the color can be discovered. In support of the ideas here advanced, it will be observed that heat, in passing through metals, actually carries something with it which affects the human system, giving headaches, etc., although the same amount of heat, passed through a transparent medium, is perfectly harmless.

In the case of a translucent substance like horn, the caloric evidently passes through pure, but the channels through which it makes its way not being straight, as in glass, the images formed on either side are broken up in the passage. Still, when the caloric has strug-gled through these tortuous channels, it is uncontam-inated by any foreign substance, and is therefore capa-ble of doing service as light, and forming new images and transmitting them to the eye.

We must now refer again to Count Rumford's doc-trine, that heat is merely a mode of motion in ordinary matter. Probably a majority of the scientists now living have adopted this idea, and, since the doctrine of the correlation of forces has been established, the belief has necessarily followed that *all* the imponderable agents are only modes of motion, either in ordinary matter or a certain ether, which really possesses all the attributes of caloric, except the indispensable one of self-repulsion. Various experiments are cited by the advocates of this theory. The one on which Count Rumford chiefly relied for the vindication of his doctrine, was the heating of a large mass of gun metal by the friction of a boring machine, which only scraped off a few chips. "Can it be possible," asked Rumford, "that so much heat can be pressed out of so small an amount of metal?" But the jar produced by the friction of boring would necessarily be transferred through the whole mass, and from every part of that mass, caloric would be shaken loose by the tremor. Sir H. Davy, too, liquefied ice by the friction of two blocks against each other, and claimed that, as ice has no attraction for the oxygen of the air with which it was in contact, it could not have attracted caloric from the air, and therefore the greater heat of the water must be simply the motion of friction in a changed state. But the true theory is that there is an enormous amount of caloric "lying around loose," dashing from place to place, and, when the structure of the ice is destroyed by friction, this caloric enters into combination with it and forms water. Count Rumford also says: "It is hardly necessary to add that anything which an *insulated* body or system of bodies, can continue to furnish without limitation cannot be a material substance." Exactly, but how can we *insulate* bodies from a substance which permeates all space, and which we know is not in the least hindered by an ordinary vacuum?

Not only is light propagated through an ordinary vacuum, but heat also permeates one as easily as it does fluids or solids. A thermometer suspended in a vacuum by a single thread of silk, responds to every fluctuation of the temperature outside. The "mode of motion," philosophers utterly fail to account for this acknowledged fact, and this brings us to the analogy with sound, on which the modern writers so much depend. Now, sound cannot be propagated through a vacuum. A bell rung in a perfect vacuum awakens absolutely no response in the organs of hearing, because there is no medium through which the motion can be propagated, and if heat and light were analogous motions the same result must follow. Mr. Grove, indeed, claims that there is a slight remnant of air, or other ordinary matter, existing in every vacuum. But,

4

if this attenuated air were the actual medium, the light and heat transmitted by means of its vibrations would affect the nerves of sight and feeling less and less, in proportion to the greater tenuity of the medium. But not only are light and heat transmitted through the infinitely small remnant of air left in the vacuum, but they are transmitted with as much, if not more, facility, than when the ordinary medium is present; consequently they cannot be, like sound, mere vibrations ot that medium.

Among other facts established by modern experiment, is the doctrine of a "mechanical equivalent of heat:" that is, the fact that a certain definite number of degrees of heat will always move the same weight the same distance. The advocates of the "mode of motion" theory claim that it is strengthened by the doctrine of mechanical equivalents, but this latter fact is even more easily explainable by the action of caloric than in any other way. What is heat, as measured by the thermometer? Simply the exercise of a force sufficient to move the mercury a certain distance. The establishment of the fact that the same amount of this force will move any other substance to a definite distance, does not change the problem in the least. Self-repulsion is the only intelligible force which can move either the mercury or its equivalent.

CHAPTER V.

THAT electricity and heat are in some manner corre-lated, has long been evident. Each generates the other, but the precise connection between them has eluded the scrutiny of the most careful experimenters. The chief difference between them is that heat is diffused from its point of origin in every direction, and conse-quently with comparative moderation and evenness; while the characteristic of electricity is its intensity, its explosiveness, its tendency to act in channels or cur-rents. If heat is simply the self-repulsive element, ca-loric, electricity is caloric confined in channels or bubbles. Air, undisturbed, is viewless and soundless; mixed in rapid motion with water and other elements, by means of friction or otherwise, it takes the form of bubbles, covered with a slight film, visible to the eye, and which burst with a perceptible noise at the slightest change of pressure. Electricity is an analogous result produced by the intermixture of caloric with other elements, usually by means of friction. The suddenness, which is an element of all electrical phenomena, is thoroughly in unison with this idea of explosiveness.

A small amount of electricity, too, is conducted thousands of miles through an iron wire; while it would require an enormous consumption of matter by combustion, to send light and heat to such a distance through the open medium of the atmosphere. If the iron is furnished with minute channels which confine the caloric, and prevent its expending its force in a lateral direction, the reason of its producing far greater results in the shape of electricity than in that of heat, which radiates in every direction, is at once apparent. The action is almost precisely like that of powder in a gun barrel. If an ounce of powder be burned in open space, its force being diffused in every direction, produces no results except in the immediate vicinity, while, if confined to a single direction, it sends the bullet to its mark a mile away. Liquids do not conduct electricity without being decomposed, because there are no channels in which the caloric can be confined.

When two bodies are brought together, both having the necessary channels, but in one of which the channels are full of caloric, and in the other comparatively empty, the self-repulsion of the element creates a current from the former to the latter, and the body which discharges the current is said to be in a state of positive electricity, while that which receives it is negative. More or less escapes in the form of radiant heat, and this alone will account for the fact stated by Poggendorf, of Berlin, that the direct discharges of an electrical machine are hotter at its positive than at its negative pole. It has been ascertained also, that the friction of precisely the same substances produces nothing but heat, while the friction of heterogeneous bodies, though also producing heat, generally evolves more or less of electricity. Why is this? If heat and electricity are both vibratory motions, arising jointly from attraction and from the repulsive motion transferred to the substances by friction, why is it that a definite amount of friction, transferred to a piece of sealing wax by another piece of sealing wax produces only radiant heat, while precisely the same amount of friction transferred by a piece of flannel produces electricity.

No reason can be given on the "mode of motion" theory. But, in truth, heat is caloric radiating freely, while electricity is caloric confined in channels or bubbles. When two precisely similar substances are rubbed together, of course the same amount of caloric is liberated from each. It meets with equal force between the sides in contact, and is radiated into the atmosphere. But, when friction is caused between two substances of different internal construction, unequal amounts of caloric are jarred loose; whichever stream is the stronger necessarily overpowers the other, and, if the substances have the necessary channels, a current is produced from the body giving out the most caloric to the one giving out the least, and the essence of electricity is caloric moving in currents.

When a current of electricity is passing through a body and the body is divided, the edges adjoining the fracture are found to be affected with different kinds of electricity; that is, the current is passing out of one part and into the other, thus: [⟶→↦] giving two kinds of electricity, but only one continuous ·current; and the same is the case where two bodies oppositely electrified are brought together. When two bodies both positively electrified are brought together, a stream of caloric is issuing from each, thus: [⇉ ⇇] so that, though the kinds of electricity are the same, there are two opposite, meeting currents.

When two neighboring bodies are negatively electrified, caloric rushes into both from the atmosphere, thus: [↙↗ ↖↗] and again, with the same kind of electricity, there are two currents moving in

opposite directions. It is easy enough to see that opposite meeting or separating currents, as in the two
last diagrams, would produce repulsion, and this we
find by experiment is actually the case. As to the
apparent attraction of substances oppositely electrified,
that is, in which there is a continuous current from
one to the other, the facts seem to be these: the positive body has a surplus of electricity whose self-repulsion continually excites it to escape, but it is partially
restrained by the pressure of caloric outside. When a
negative body is brought close to it, there is no pressure from that side, because there the caloric is going
the other way; therefore the pressure on the other
sides pushes the two bodies together. In case of light
bodies, silk, pith balls, etc., this pressure is sufficient
to overcome gravitation and produce apparent attraction; the workings of this same force on certain heavier
bodies is called magnetism, and will be dwelt upon
presently. When the approach of a negative body to a
positive one takes away the restraining pressure on that
side, the confined caloric frequently leaps out to a consid-
erable distance. When it bursts from one cloud to another, or from a cloud to the earth, it is called lightning.

Experiments also show that electricity can hardly be
made to cross a vacuum, and therefore Mr. Grove con-

cludes that electricity being a mode of motion of ordinary matter, is interrupted by a space from which such matter is absent. But we know that light and heat are not interrupted by a vacuum, consequently we must look for some other cause. This cause we find in the fact that electricity is caloric acting in channels, and, as a vacuum furnishes no outside compression, the caloric on entering it is dissipated as electricity, though still existing as radiant heat. In accordance with the above views we find, by the experiment of Matteucci, that bismuth conducts electricity better in the direction of the planes of cleavage, than across them. Why? Evidently because there are channels in that direction for it to run in.

The workings of magnetism are perhaps the most mysterious, to ordinary apprehensions, of all the different manifestations of natural power with which we are acquainted. That a bit of iron or lodestone should overcome the power of gravitation, which seems so universal, and draw upward other pieces of the same metal; that, when poised, it should invariably assume a north and south direction, seems at first, scarcely less than miraculous, and might well excuse the phantasy of the old writers who endowed the magnet with a living soul. As in the case of suction, gravitation, lightning, etc., the first attempt to explain it was by attraction. The lodestone was said to have in it an extraordinary attractive power, sufficient to overcome that of gravitation, and there was supposed to be an enormously large magnet near the north pole, which attracted all the other magnets toward it.

Advancing science, however, has established the correlation of magnetism and electricity, until it is now thoroughly understood that the former is, in some way, produced by electric currents. The general opinion, however, of English scientists is, that electric currents, like other manifestations of force, are only a species of "molecular polarization," and that magnetism comes under the same head. But this same "molecular polarization" depends on the attraction and repulsion of ordinary matter, and, as attraction is impossible, we must look to the action of a self-repellent fluid on inert matter. In trying to find how the currents of this fluid affect the lodestone or iron so as to produce the phenomena of magnetism, we shall proceed most securely by observing the method by which artificial magnets are made. This is done by passing a heavy electric current *through* (in ordinary phrase) a bar of iron. If soft iron be used, a temporary magnetization is the result; if hard, a permanent one. Though Mr. Grove's ideas of "molecular polarization" are extremely vague, yet his explanation of the manner in which electric currents act in forming a magnet is so clear, that I cannot do better than to quote it:

"Suppose a number of wind vanes, say of the shape of arrows, with the spindles on which they revolve arranged in a row, but the vanes pointing in different directions; a wind blowing from the same point with a uniform velocity will instantly arrange these vanes in a definite direction, the arrow-heads or narrow parts pointing one way, the swallow-tails or broad parts another. If they are delicately suspended, a very gentle breeze will so arrange them, and a gentle breeze will again deflect them; or if the wind cease and they have been originally subject to other forces, such as gravity from unequal suspension, they will return to irregular positions, themselves creating a slight breeze by their return. Such a state of things will represent the state of the molecules of soft iron; electricity acting on them—not indeed in straight lines but in a definite direction—produces a polar arrangement, which they will lose as soon as the dynamic inducing force is removed.

"Let us now suppose the vanes, instead of turning easily, to be more stiffly fixed to the axles, so as to be turned with difficulty; it will require a stronger wind to move them and arrange them definitely, but when so arranged they will retain their position; and should a gentle breeze spring up in another direction, it will not alter their position, but will itself be definitely deflected. Should the conditions of force and stability be intermediate, both the breeze and the vanes will be slightly deflected; or if there be no breeze and the spindles be all moved in any direction, preserving their linear relation, they will themselves create a breeze. Thus it is with the molecules of hard iron or steel in permanent magnets; they are polarized with greater difficulty, but, when so polarized, they cannot be affected by a feeble current of electricity. Again, if the magnets be moved, they themselves originate a current of electricity, and lastly, the magnetic polarity and the electric current may be both mutually affected."

This shows as clearly as possible that magnetism is not an inherent attribute of the magnetic substance, but is the result of the action of electric currents upon it. As, moreover, we cannot produce an artificial magnet except by subjecting it to the action of an electric current, it is, in the absence of any other known method, but fair to presume that the natural magnet is formed in the same way, except that in the latter case the fiery agent comes fresh from Nature's laboratory, and in the former from an electric battery. As these currents cannot be caused by an attraction which has been demonstrated impossible, they must be referred to the self-repulsion of an actual fluid which we call caloric.

The next inquiry is as to what direction the currents take in regard to the iron, in order to make it magnetic, and again the best method of answering it is to look at the means used to produce artificial magnetism. One method is to send a strong current of electricity *through* (as is generally supposed,) a piece of iron; the other is to send one *around* it by means of a spiral coil of wire. In the latter case, we know that the current does go around the iron, while in the former, though we use the word *through*, yet we only know that the electricity, in some way, gets from one end of the iron to the other; it may go through, or it may, as in the former case, go spirally about it. All the positive evidence is in favor of the spiral direction of the currents, because we only know with certainty that we can produce magnetism in this way, and, as precisely the same result is obtained by the transmission of an ordinary electric discharge along the magnet, it is most philosophical to suppose that these identical results have been produced by like causes, and that the current, in all cases, passes around the iron spirally, in order to produce magnetism.

Besides, it cannot be that the mere conduction of an electric current produces magnetism in heavy bodies, for, if it did, silver, which has eight times the conductive power of iron, would be, in that proportion, more magnetic than the latter substance, while, in fact, it has no magnetic properties whatever. The theory of Ampere, which is sustained by the experiments of Oersted, is evidently the true one; the channels which conduct caloric in iron and a few other substances are different from those in most electric conductors, running spirally in such manner that the current encloses the iron. When the ends of two pieces of iron are brought together, in one of which the current is circulating from the end towards the center, and in the other from the center toward the end, they are said to be endowed with opposite polarity, but there is a continuous current which results in apparent attraction; while, if both currents are going towards the center, or towards the end, there is similar polarity, but opposite movement of the currents, producing repulsion. That this is the true reason for the apparent attraction of magnetism, is also shown by experiments recorded in many works on electricity, proving that there is a repulsive action set up between two wires along which electrical currents are moving in opposite directions, but that, when the currents move in the same direction along each wire, they produce attraction.

But, though Ampere has correctly stated the relation of the currents from which attraction results, yet he fails to show how it is that the flow of a continuous current around two magnetic substances causes them to approach each other. In fact, there can be no such thing as attraction in nature, and in this case, as in all others, apparent attraction is due to outside pressure. The whirling currents enclose the iron, cutting off all pressure from the sides, and creating a kind of eddy, while the ever active fluid on the outside presses against the outermost ends and drives the magnetized bodies together.

A further proof that the magnetic current always acts in a spiral direction, is found in the fact that after it leaves the conductor it still maintains its spirality. This is plainly shown by a little experiment recorded by Grove, when speaking of light. When a ray of polarized light is submitted to magnetic action, it is twisted, "as if," in the words of Mr. G., "a card had been driven forcibly through a grooved rifle barrel." Connecting this last fact with the production of magnetism by electricity passing around the magnet, and it is difficult to see how any one can resist the conclusion that magnetism results from the formation of a kind of eddy within the spirally revolving currents of electricity or caloric.

These spiral channels in the magnetic substance are evidently opened by a blast of electricity, permanently in the case of hard iron, temporarily in that of soft, as explained in the quotation from Grove. The channels once opened, the rays of caloric everywhere flying through space strike into and pass through them, until some new disturbing clause closes them up. This theory also accounts for the remarkable results of an experiment recorded by Professor Tyndall. After magnetizing a part of a link of a chain cable by means of a coil of wire, around it, he suspended a piece of silver between the poles by a twisted cord, and found that the thread was prevented from untwisting, when the current was in circulation, by a certain viscousness or thickness of the atmosphere; "as if," says the professor," the silver piece were struggling in treacle." When he undertook to penetrate this strange invisible obstruction, it was like sawing through soft cheese. This can easily be accounted for on the theory of spiral currents, and on no other. The two eddies at the poles

of the magnet exercise a *repulsive* power on the dia-
magnetic silver or copper, holding it in its place and
preventing the string from untwisting.   It will perhaps
be clearer, if we imagine two big pairs of bellows sending two very strong currents of air exactly against
each other, and a broad thin object placed between
them; would there not be some difficulty in twisting it
around? Not only silver, but all other non-magnetic substances are feebly repelled by a magnet, showing clearly
that magnetism is originally a repulsive power, and
only becomes attractive when the spiral channels of
magnetic substances create a vortex in its lightning-
like waves.

If the electric current went straight through the
magnetic substance, whatever force it exerted would
tend to bring the body in line with the current, but, as
it goes round and round it, the repulsive force of the
fluid compels the iron to lie lengthwise of the vortex
in which it is enveloped, but crosswise of the general
course of the current. This last fact, after numerous
investigations, is set forth as a law by Ampere. A
homely illustration is seen when a stick is thrown into
an eddy.   The faster the water flows the more nearly
the stick assumes a perpendicular position, that is, a
position at right angles with the current. If the water

went as fast as lightning, the stick would certainly be entirely perpendicular. For this reason the electric currents around the earth, when they strike a magnet and begin to circle around it, compel it into a position at right angles to their course, or nearly north and south.

5

CHAPTER VI.

THESE are the most universal of all the affections of matter, and are the ones in which the workings of attraction appear at first sight to be the most complete. Cohesion designates the power which binds particles of the same nature in a single mass, while chemical affinity is the name applied to that mysterious force which compels different primary substances to unite together in certain definite proportions, thus forming other substances, with qualities often entirely unlike those of their ingredients. This certainly looks like attraction, but so does the forcing of water up a tube from which the air has been exhausted, and so does the motion of iron towards a magnet. As, however, it is impossible that there should be any such thing as attraction, let us see if the phenomena of chemical affinity and cohesion cannot be explained by the law of self-repulsion.

Caloric is locked up in all substances, between the atoms; the pressure of caloric from the outside holds these atoms together, and the hardness of the substance is in proportion to the closeness with which the atoms are fitted to each other. The same amount of caloric always expands with equal force, and, in any substance, the force of the enclosed caloric must be less than the outside pressure, or else the body would be disintegrated by its own internal power. When atoms of the same kind are in a mass together, the walls of the cells are of the same strength; consequently no internal change takes place, and the external pressure (not merely of the atmosphere, but of free caloric,) drives the atoms together with a force which is commonly called cohesion.

When, however, different substances are intimately mingled, the walls of their cells are of different degrees of strength. If two adjoining cells be represented by "a" and "b," thus: ⓐⓑ it is evident that, so long as the pressure on the central wall is equal from each side, it will not be torn down, but if, in place of "b," we substitute another cell, "c," in which the caloric is more compressed than in "a," it will overcome the pressure of "a" and produce a rupture. If "c" contains less force than "a," it will be overcome by the latter and the same result will follow. When different substances are mixed in certain definite proportions, determined by the character of both, there will be enough of these *weak points* in the mixture to produce a mingling of the whole in a homogeneous mass, which will have a new set of cells, and will affect all our senses differently from either of the component substances; and this effect of the self-repulsion of caloric is called chemical affinity.

Equality in the amount of force locked up in the different cells is fatal to chemical affinity, but is highly promotive of cohesion. It is like leaning together a dozen sticks to support each other; if they are all heavy, or all light, the equal balance between them will prevent their being borne to the earth by gravitation, but if we put a few very light ones, or a few very heavy ones, on one side, either the others will overbear them or they will overbear the others, and the whole will fall into a single, confused heap. But the change would have to be made in definite proportions. If we only substituted one light stick, the balance would probably still be near enough equal to prevent a collapse; while if we substituted eleven, a new equality would be introduced with the same result. Coarse as this illustration is, we believe it will give a pretty fair idea of the workings of chemical affinity.

Often, when different substances are intermingled, the

enclosed caloric is not sufficient to burst the walls until a sudden shock is given, generally by electricity, when, the breaking process being once started, the old walls are instantly destroyed, precisely like a child's house of cards, which falls at the slightest disturbance. The diamond, so far as chemical analysis can determine, is pure carbon; charcoal is the same; and, if simple attraction caused the cohesiveness of both, the substance being the same, there is no reason why the cohesiveness of the two should not be equal. As, however, cohesion depends, not on internal attraction, but on external pressure, we can easily understand how the arrangement of the particles would cause greater hardness in the substance, in proportion as the closeness of their fit prevented the caloric from penetrating between them.

Whenever the proportion of caloric to ordinary matter in any substance is large, the application of flame breaks down some of the walls, and then the escaping caloric, by its violence, tears down others, and so on till the structure of the substance is destroyed by *combustion*. Generally, this process is assisted by the presence of oxygen, an elastic gas, containing a large proportion of caloric, which escapes readily and aids in breaking down the cellular walls of the combustible.

Sometimes, in the operations of nature or art, an enormous amount of caloric becomes locked up in weak cells, but is so precisely balanced by the outside pressure that it does not escape until, by the addition of heat, the equilibrium is destroyed; then all the enclosed caloric rushes forth at once, and *explosion* is the consequence.

CHAPTER VII.

ADHESION is one of the simplest forms of outside
pressure fastening substances together, and producing
apparent attraction. One of its most familiar examples
is seen when a boy fits a piece of wet leather neatly
to a smooth stone, and lifts them both together. In
this case the air is excluded, and the pressure of that
element from the outside holds the leather to the stone.
A similar result is attained when we fasten two pieces
of wood together with glue, or two pieces of paper
with mucilage. In these and all similar cases, the
whole secret of adhesion is to make a mixture through
which but little caloric will pass. Then, with this ever
moving, self-repellent fluid working against the outside,
without being able to penetrate between the two substances, they can only be torn apart by muscular force,
which last is but another exemplification of the power
of caloric, as will be shown in the next chapter.

ELASTICITY is still another species of force, which can be traced directly to the self-repulsion of caloric. When that element is locked up in the minute globules of different substances, being equally self-repellent in every direction, it has a tendency to force the inside of each cell into a spherical form. Sometimes the small amount of the enclosed caloric allows the atoms to be arranged in their original cubical form, thus producing crystals, and, in proportion as this is the case, the substance is without elasticity. Extremely cold substances are never elastic, of which fact the destruction of the elasticity of iron by extreme cold is a familiar proof. When, however, these spherical cells are very numerous in a substance, it is highly elastic, and its elasticity acts in this way: if the substance is bent, the cells on the side toward which it is bent are flattened, while those on

the other side are elongated, thus:  When

the bending power is withdrawn, the caloric again forces the cells into a spherical form, because that is the shape in which it has most room to exert its self-repellent power; and thus one side is pulled, and the other is pushed, back to its original position.

Most substances only show their elasticity when bent in a lateral direction, but some articles, such as India-rubber, can be elongated with remarkable ease. In this case all the cells are drawn out, thus: but, as soon as the stretching force is withdrawn, they resume their globular form, thus:

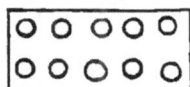

CAPILLARY ATTRACTION is another mode of force arising from the self-repulsion of caloric, which has been explained in Chapter III.

THE TIDAL WAVE is another exemplification of apparent attraction, but of actual repulsion. Force is acting on the globe from every side; the moon cuts off some of this force from the portion of the globe opposite to it, and the outside pressure crowds the water up under the moon. This is plain enough, but why is the water at the same time lifted upon the opposite side of the earth from the moon? Herschel acknowledged the apparent absurdity of attributing this phenomenon to the attraction of the moon, and attempted to explain it by a calculation of the difference in attraction between one side of the moon and the other, which appears to me at once unintelligible and inadequate.

It can, however, be easily explained on the theory of outside pressure. The sea is not a mere body of inert particles, but an elastic substance, covering the greater part of the earth; that is, a substance in which caloric is enclosed, and which expands in length when a pressure is applied at the sides, and the reverse. Let us represent this elastic substance, sustaining pressure from all sides, thus: →○← Now remove the force A, (as is done by the interposition of the moon.) Now there will be a double pressure in the direction D C, and only a single pressure in the direction A B. The result is that each elastic particle has a tendency to be elongated in the direction transverse to the greatest pressure, thus: ⬭← and the joint action of billions of these particles produces the tides on both sides of the earth at once, in a line *transverse to the line of greatest pressure.*

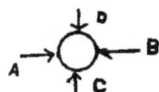

CHAPTER VIII.

THE weight of opinion in former times seems to have been that the heat, which all could see was necessary to animal and vegetable life, acted rather as an oil to lubricate, or a stimulus to set in motion, the vital machinery, than as a force of itself, and the living body was supposed to exert some kind of a power of its own, usually called the vital principle. A long train of experiments by different modern philosophers has suggested, with more or less distinctness, the identity of heat with this vital force, which builds up the vegetable kingdom and operates on the blood and muscles of the animal.

Among the most distinguished of the writers on this subject is Dr. Carpenter, who, in an essay on the Correlation of Physical and Vital Forces, shows conclusively that heat is the force which drives forward both plants and animals in their growth and action; that, in regard to the former, the necessary heat is evolved from earth, air and water, while in the case of animals, it is elaborated principally from the food they eat, and that it then performs mechanical labor, very much as it does in a steam engine.

He divides organic action into two parts, the force which directly produces growth or action, and the *germ power*, or peculiar principle, which is only directory. By means of this directive agency, the heat acts in particular channels, producing the millions on millions of varieties of plants and animals throughout the world. As to the nature of this germ power or directive agency, we are still in the dark, and Agassiz mentions this development of different animals from apparently similar germs, as one of the strongest evidences of the interposition of a creative will in the affairs of earth.

But, though we cannot explain the origin of the directive channels through which the vital force acts, we can see that the power itself is heat, both in plants and animals. The greater the heat, the greater the force and activity of the organization. Without sufficient heat, a seed may remain undeveloped for an unknown length of time, yet retaining all its germinal nature, ready to give direction to the force of heat whenever it shall be applied. The process of malting depends entirely on the certainty of producing a definite advancement of germination in the seed by a definite increase of heat.

How heat can produce the different organic movements, Dr. C. is unable to explain, and indeed, if we adopt the doctrine of Tyndall, and others, that heat is but a mode of motion, and not due to the properties of an actual fluid, it is as difficult to understand how the single force, attraction, can produce the circulation of the blood as the circulation of the solar system. But the action of the self-repellent fluid, caloric, makes all plain, so far as the motion alone is concerned. Entering into the plant through the roots and leaves, the caloric, by its self-repulsion, drives the sap through the veins, and pushes the solid substances derived from the earth up where they will strengthen the trunk. By means of that destruction of cellular walls which we call chemical affinity, it produces out of different ingredients new and more plastic compounds, and these it pushes into their places, where, by the further action of the self-repulsion of caloric, that element is driven out of the fluids, leaving them deposited as solids, so that, in the words of Dr. Carpenter, "the final cause or purpose of the whole vital activity of the plant, so far as the *individual* is concerned, is an indefinite extension of the dense, woody, almost inert, but permanent portions of the fabric, by the successional development, decay and renewal of the soft, active, and transitory cellular parenchyma." We have seen, in the case of magnetism, that a certain arrangement of the inert particles of a substance, a certain medium between total obstruction and entire liberty to the currents of caloric, causes them to move in a spiral direction, and the same law, whose operation is seen in magnetism and the action of polarized light on certain substances, is also discovered in the spiral course of vines and twining plants, as the caloric in their veins overcomes the inertia in their woody matter.

The correlation between heat and animal life is still more evident. Caloric escapes from the food taken into the stomach. It is self-repellent; it must do something; if it be allowed any space at all, it must change its position with lightning-like rapidity. It pushes its way into the chyle, and thence into the blood. Still it is self-repellent, and, by its own inherent force, it drives the blood before it to the extremities and back again to the heart, always following the direction marked out by the unexplained germ power. Then it escapes through the pores of the skin, or otherwise, giving way to a fresh supply of the same self-repellent element.

The greater the amount of caloric the more easily it carries the inert matter with it; consequently, in strict accordance with this theory, the hot blood rushes along with torrent-like rapidity, the cold blood creeps with slow and lagging motion. So far as the circulation of the blood is concerned, the doctrine of Harvey has been universally received, that the contraction of the walls of the heart propels the blood through the arterial tubes and back through the veins, the direction of its movements being insured by a proper arrangement of the valves. On this subject the eminent physiologist, Professor Draper, well remarks:

"But, when comparative anatomy and physiological botany were more extensively cultivated, it was seen that this doctrine is insufficient; for the unity of Nature forbids us to believe that nutritious juices are circulated in different tribes of life by different forces. And, though it may be true that the contractions of that central impelling mechanism (the heart,) regulate the circulation in the organisms which have a heart, what is to be made of those countless numbers which have none?. In this group we find the whole vegetable creation, and a majority of the animal."

And again: "The circulation of sap in plants, and of blood through many of the lower animals which have no heart, is active during summer, checked by cold nights, and wholly arrested during winter."

The pulsations of the heart are rather the effect than the cause of the flow of the blood, the latter action having for final cause the self-repulsion of caloric. "Forty-five minutes," says Hunter, "after the vessels of the neck of a tortoise had been divided, the pulsations of its heart were augmented from nineteen per minute to thirty-five, by increasing the temperature from that of the room to 90°."

Numerous modern experiments confirm this theory, one of which, by Herr S. L. Schenck, is as follows: "Examining the heart of the chick in the egg of the fowl, he discovered that its movements are at first quite independent of the central nervous system, and may be regarded as simple contractions of the protoplasm. When the heat is removed it still beats, if maintained at a temperature of 34° to 36° centigrade. The most powerful microscope fails to show any trace of nervous ganglion; hence he concludes that the contractions of the heart are due simply to the action of heat on the protoplasm."

Yes, *heat, heat, heat,* with self-repulsion as its attribute, is the universal material cause of motion, whether in the orbits of planets or the veins of a tortoise.

Again I quote from Dr. Carpenter: "That the developmental force which occasions the evolution of the germ in the higher vertebrates is really supplied by the heat to which the ovum is subjected, may be regarded as a fact established beyond all question. In frogs and other amphibia, which have no special means of imparting a high temperature to their eggs, the rate of development (which, in the early stages, can be readily determined with great exactness,) is entirely governed by the degree of warmth to which the ovum is subjected." The same is true (though perhaps it cannot be so definitely measured,) of the eggs of serpents and birds, and also of the ova of mammalia, which receive the heat necessary for their development before their separation from the mother.

Another remarkable illustration of the power of heat, or caloric, is found in the phenomena of reparation. Certain cold-blooded animals are so constructed that, when a portion of their organism is cut off from the rest, even to the extent of a whole limb, a new one will grow out in its place, as the hair and nails grow on the human frame. But, says Dr. Carpenter: "Not only can the rate at which they take place be experimentally shown to bear a direct relation to the temperature at which these animals are subjected, but it

6

has been ascertained that any extraordinary act of reparation, such as the reproduction of a limb in the salamander, will only be performed under the influence of a temperature much higher than that required for the maintenance of the ordinary vital activity."

Freezing and burning of the flesh are the result of apparently opposite causes, and, at first, produce opposite sensations on the nerves; yet their physical effects on the flesh are nearly the same. Why? Because both freezing and burning result in tearing down the cellular walls of the flesh. These cells, in their normal condition, contain caloric in nearly equal amounts. When a heated body is placed against the skin, a larger amount of caloric than usual finds its way into the outer cells. Adjoining these are the interior cells, much less distended by heat, and, where the difference is very great, the expansiveness of caloric in the distended cells breaks down the filmy tissues between them and the others, producing a blister. When a cold substance is placed against the outside of the flesh, the caloric leaves the outer cells, and the expansiveness of that element confined in the others also breaks down the tissues, and produces the same result as in case of a surplus. The operation of the caloric in tearing down the tissues is about the same as in chemical affinity, only in that case the adjoining cells are made weaker or stronger by the substitution of others belonging to different substances, and, in this case, by exhausting or overcharging those that already lie side by side.

Still more evident is the operation of caloric in producing muscular motion, which Dr. Carpenter does not include among the correlated manifestations, but mentions as "the most important consideration of all, namely: the source of that contractile power which the living muscle possesses, but which the dead muscle, though having the same chemical composition, is utterly incapable of exerting."

In producing this contraction of the muscles, there are two steps: the mind (in man—the instinct in lower animals,) acts by means of the brain on the nerves, and these act on the muscles. The first of these steps we must as yet leave among the mysteries, but the action of the nerves on the muscles, producing the contraction of the latter, is due entirely to the self-repulsion of caloric. That element, elaborated from food and air, is stored away in the different tissues of the body. All muscles are of a long oval shape, with a swell in the center, so that their expansion laterally produces contraction in length. By some, as yet, unexplained power, the mind opens the infinitesimal valves of the nerves so as to admit caloric into the empty muscles. Instantly its self-repulsion drives the sides of the muscles outward, and that pulls the end upward, thus moving the bone to which it is attached.

Numerous experiments show the truth of this theory. Sending an electric current aldng the nerves of an animal in a direction from the brain, produces contraction of the muscles in which these nerves are inserted; that is to say, it expands them in breadth but contracts them in length. If the leg of a frog be curled up into the smallest possible space, an electric current will cause it to stretch out to its greatest length, as if the animal's instinct had ordered it to hop its best. The case is precisely the same as if an elliptical India-rubber bag had been curled up, and then expanded by the admission of air.

Still further evidence on the point is furnished by the fact that successive slices of animal muscle exhibit different kinds of electricity, the inside being positive, the outside negative; that is, the caloric is slowly making its way through the muscular tissues from the inside outward. As it escapes, its place must be supplied from other parts of the body, and thus the whole system may be exhausted (producing the sensation of fatigue,) by the long continued exertion of one arm.

Surprise has often been expressed at the enormous power exerted by the muscle in lifting the lever-like bone of the arm, the muscle being attached close to the fulcrum, while the weight is placed on the further end, thus greatly increasing the force necessary to raise it. It is evident that the same substance whose self-repulsion produces motion across the illimitable fields of space at the rate of millions of miles per minute, and rolls planets and satellites along their mighty orbits, must, when admitted into the muscles, swell them in width and contract them in length with tremendous force, and that is the only material power with which we are acquainted that can produce such results.

But the strongest evidence of the true nature of muscular action is to be found in the examination of the muscles themselves. We see at once that when we lift a heavy weight the muscles are swollen in thickness, and, if we dissect them, we find that the separate fibrils are swollen by apparent contraction. In its construction (see Draper's Physiology,) each fibril is a series of cells, the diameter of which varies as the muscle is in the contractile or relaxed state, but which may be taken on an average at the one-thousandth part of an inch. When the fibril is in a relaxed state, the longest axis of each cell coincides with the length of the fibril, but,

when contractions occur, this axis shortens, the transverse axis broadens, and a shortening of the entire
fibril is the result. The most accurate tests that can
be applied show no diminution in the volume of a
muscle when contracted, and Draper, while objecting to
the kind of test used, says that if the instruments were
delicate enough, they could probably show the "*preposterous* result of an increase of volume by contraction."
Certainly, if it *were* contraction, such a result would be
preposterous, but it is really expansion. The Professor
also says, it is well known that the muscle becomes
warmer when contracting; this strengthens the argu-
ment in favor of the theory that the contraction of
muscles in length is the result of their expansion in
width by the caloric admitted into them. Again, the
same author, in speaking of the knots or swellings in
the nerves, called ganglia, says: "Ganglia permit the
influence passing along the nervous cords to escape
therefrom into new channels, and also retain and store
up nervous power. They become, therefore, magazines
of force, and are hence capable of sustaining rhythmic
movements." It cannot be that a mere abstract attractive power could be retained in a knot or swelling.
Certainly not. Something which requires space for its
operation must be confined in these "magazines of
force," and caloric is the only element which, in its own
self-repulsion, contains a permanent fountain of force.

The power of electricity to produce longitudinal contraction (lateral expansion) of the muscles, has already
been alluded to. If heat and electricity are, as is generally believed, identical forces, then we would natu-
rally suppose that heat, if properly applied, would produce the same result, and this idea we find to be borne
out by experiment. Hunter has left this statement:
" I took three pieces of muscle from the neck of a sheep
just bled to death, each of them an inch and a half in
length, and put them into separate vessels of water, at
temperatures of 80°, 110°, and 120°, when the one in
the first vessel contracted one-third of an inch in three
minutes; the second one-half an inch; and the third
three-fourths in one minute and a half." Can this pos-
sibly be anything else than the self-repulsion of caloric
in the cells of the muscles ?

Forms of speech often gain wide currency among the
people which cannot, at the time of their origin, be
proven to be strictly true, but which have nevertheless
been selected with apparently instinctive insight to express the ideas which subsequent investigations demonstrate to be correct. Thus, housewives frequently say
that when flour is ground too fine it "takes the life
out of it." This has generally been considered a tolerably fair expression of the fact that the nourishing
value of the flour was diminished, and nothing more,
but it would be difficult to give a more literal state-

ment of the truth. Caloric is shut up in the flour. When the latter is reduced to excessively small particles, the former escapes; consequently there is a less amount to be set free by the process of digestion; to enter into the blood, the nerves, the muscles, and there by its self-repellent power to produce that motion for which the word life is but another name.

So, too, the expressive word "fire-water," which the Indians apply to intoxicating liquors, conveys a literal truth. It is the fire, the caloric, shut up in the liquor, which escapes from it into the veins and nerves. By its self-repellent power it gives greater rapidity to the blood, and for a while distends the muscles with more than ordinary force. But, in its fiery course it disorganizes the delicate tissues, so that the natural amount of caloric evolved from food is not properly retained, and consequent weakness is the result. The will of man, too, is capable of controlling the nerves when only a moderate amount of caloric courses through them, but, when the enormous amount evolved from alcohol is added, all control over the tiny valves which regulate its flow is lost, and a hundred involuntary movements betray the ungoverned action of the self-repellent fluid.

CHAPTER IX.

THE impossibility that attraction should be the basis of natural phenomena has probably suggested itself to many minds. Newton saw its absurdity as mentioned in Chapter I. [Since writing the foregoing, I find that Newton actually expressed the opinion that gravitation was caused by a subtle ether pervading all space, only stopping short at the mode of its operation.] Faraday has demonstrated its inconsistency with the law of the conservation of force. But none appear to have enunciated the actual law. The nearest approach to the truth, which I have as yet discovered, is in a few incidental notes to a work entitled " A System of Logic," by a Mr. Gregory, which has come under my notice since I began this work. Mr. G. says: " That the sun really attracts or draws the earth or any other planet, is an absolute impossibility, because it has no hold on them, and consequently it can no more draw them than it can empty space. * * * The proofs given of this law (gravitation,) are quite fallacious, so far as they attempt to show that there is any real attraction, their authors having overlooked the fact that all the phenomena may result from a compulsive instead of an attractive force. * * * Attraction, without connection, is a manifest impossibility. Another difficulty in the way of attraction is that the bodies are inanimate, and therefore it is evidently as impossible for them to move either themselves or other bodies as it is for a rock to move itself from one mountain to another."

Mr. G., however, gives no decided opinion as to the true cause, but supposes it may be from *waves* of gravitation, acting from the outside, and says: "According to this view, gravitation is a compulsive and not an attractive force, as it is constantly termed, or in other words it is a *pushing* and not a *pulling* force."

This expresses the truth very correctly as to the mode of action, but where do the *waves* come from? It is idle to refer motion to some other motion; we must seek some permanent attribute of matter on which to found it, and the self-repulsion of caloric is the only one which will produce the known result. Dr. S. S. Metcalfe, of Transylvania College, Kentucky, author of a voluminous work on caloric, grasped the fact that the self-repulsion of that element was a great agent in producing motion, but he spoiled his theory by adding that caloric, while repelling its own particles, attracted those of ordinary matter; both attraction and repulsion being "inversely as the square of the distance." This idea is subject to all the objections pointed out by Faraday, and is incompatible with the fact that extreme coldness and great solidity usually accompany each other.

Though Prof. Tyndall is wrong in denying the existence of caloric, yet it is impossible to demonstrate the conservation and correlation of forces as thoroughly as he and other masters of modern science have done, without seeing that heat is the force underlying all others, and showing itself from under a thousand disguises. In one of the lectures of the great scientist just named, the omnipresent power of heat is depicted with a vigor and eloquence which, as I cannot equal, I will quote:

"Grand, however, and marvelous as are those questions regarding the physical constitution of the sun, they are but a portion of the wonders connected with our luminary. His relationship to life is yet to be referred to. The earth's atmosphere contains carbonic acid, and the earth's surface bears living plants; the former is the nutriment of the latter. The plant apparently seizes the combined carbon and oxygen; tears them asunder, storing the carbon and letting the oxygen go free. By no special force, different in quality from other forces, do plants exercise this power—the real magician here is the sun. We have seen in former lectures how heat is consumed in forcing asunder the atoms and molecules of solids and liquids, converting itself into potential energy, which reappeared as heat when the attraction of the separated atoms was again allowed to come into play. Precisely the same considerations which we then applied to heat we have now to apply to light, for it is at the expense of the solar light that the decomposition of the carbonic acid is effected. Without the sun the reduction cannot take place, and an amount of sunlight is consumed exactly equivalent to the molecular work accomplished. Thus trees are formed, thus the meadows grow, thus the flowers bloom. Let the solar rays fall upon a surface of sand, the sand is heated and finally radiates away as much as it receives; let the same rays fall upon a forest, the quantity of heat given back is less than that received, for the energy of a portion of the sunbeams is invested in the building of the trees. I have a bundle of cotton which I ignite; it bursts into flame and yields a definite amount of heat; precisely that amount of heat was abstracted from the sun in order to form that bit of cotton. This is a representative case; every tree, plant and flower grows and flourishes by the grace and bounty of the sun.

"But we cannot stop at vegetable life, for this is the source, mediate or immediate, of all animal life. In the animal body vegetable substances are brought again into contact with their beloved oxygen, and they burn within us as a fire burns in a grate. This is the source of all animal power, and the forces in play are the same in kind as those which operate in inorganic nature. In the plant the clock is wound up, in the animal it runs down. In the plant the atoms are separated, in the animal they recombine. And, as surely as the force which moves a clock's hands is derived from the arm which winds up the clock, so surely is all terrestrial power drawn from the sun.

"Leaving out of account the eruptions of volcanoes and the ebb and flow of the tides, every mechanical action on the earth's surface, organic and inorganic, vital and physical, is produced by the sun. His warmth keeps the sea liquid and the atmosphere a gas, and all the storms which agitate both are blown by the mechanical force of the sun. He lifts the rivers and the glaciers up to the mountains, and thus the cataract and the avalanche shoot with an energy derived imme-

diately from him. Thunder and lightning are also his
transmuted strength. Every fire that burns and every
flame that glows dispenses light and heat which originally belonged to the sun.   In these days, unhappily,
the news of battle is familiar to us, but every shock
and every charge is an application or misapplication of
the mechanical force of the sun.   He blows the trumpet,
he urges the projectile, he bursts the bomb.   And
remember this is not poetry, but rigid mechanical truth.
He rears, as I have said, the whole vegetable world, and
through it the animal; the lilies of the field are his
workmanship, the verdure of the meadows, and the
cattle upon a thousand hills.   He forms the muscle, he
urges the blood, he builds the brain.   His fleetness is
in the lion's foot; he springs in the panther, he soars
in the eagle, he slides in the snake.   He builds the
forest, and hews it down; the power which raised the
tree, and which wields the axe, being one and the same.
The clover sprouts and blossoms, and the scythe of the
mower swings by the operation of the same force.

"The sun digs the ore from our mines, he rolls the
iron, he rivets the plates, he boils the water, he draws
the train.   He not only grows the cotton, but he spins
the fibre and weaves the web.   There is not a hammer
raised, a wheel turned, or a shuttle thrown, that is not

raised and turned and thrown by the sun. His energy is poured freely into space, but our world is a halting place where this energy is conditioned. Here the Proteus works his spells; *the self-same essence takes a million shapes and hues, and finally dissolves into its primitive and almost formless form.* The sun comes to us as heat, he quits us as heat, and between his entrance and departure the multiform powers of our globe appear. They are all special forms of solar power; the molds into which his strength is temporarily poured in passing from its source through infinitude."

Striking out all reference to attraction, the foregoing is true, except that in place of the sun, which is the mere organ of action, we should substitute the self-repellent element, caloric, or fire, eternally flashing through all space, leaving the sun only to fly across immeasurable fields until it strikes some planet or star; reflected, refracted, driven into the bowels of the earth, or Jupiter or Sirius, by the force, derived from its own self-repulsion, with which it strikes those bodies; driven out again by that same self-repulsion, eternally producing motion, and returning perhaps, after millions of years, to the sun, whose power it continually supplies. In short, we must substitute for the sun, which is always millions of miles away, the caloric which is here in our midst one minute, and in the next instant may be aiding some old lady of Saturn to thread her needle, or thawing a snow bank on the mountains of Uranus.

Instead of overstating the effect of this wonderful force, the learned Professor does not make a large enough list of its manifestations. It does not merely lift up the water, leaving attraction to draw it down, but the same power which raises the fluid drives it back to earth, and drives the earth itself along its orbit.

When a man pumps water out of a well into a trough, where the cold first freezes it and then the sun converts it into vapor, *it seems* as if the man *contracts* his muscles, thereby working the pump, which *sucks* the water up till it issues from the spout, when the earth *attracts* it towards its center, and when it is frozen in the trough it seems as if it is held together by the *attraction* of cohesion, and that the sun destroys that cohesion by means of chemical *affinity* and *attracts* the vapor upwards. But, in fact, the man admits caloric into his muscles, where by its self-repulsion it *expands* them in width, thus drawing up the ends, lifting the bones, and working the pump. When a vacuum is formed, the *pressure* of the air on the outside *drives* the water up till it runs from the spout, when the concentrating rays of caloric *drive* it downward. Then the caloric escapes into the colder atmosphere, and the particles fit so closely that the caloric on the outside *presses* them in a solid mass. When a still greater amount of caloric is driven against this mass from the sun, it enters between the particles, *pushes* them apart by its *expansive* power, and then by the same power is *driven* upward, carrying with it the small particles we call vapor.

7

CHAPTER X.

CALORIC—Is a subtle, eternal, self-repellent fluid, extending through all space.

Attraction, (apparent,)—Is the result of screening a substance from pressure on one side, leaving it to be driven towards the screen by the pressure from the other side.

Heat—Is caloric acting on the sense of feeling.

Light—Is caloric acting on the sense of sight.

Gravitation—Is the driving of bodies towards the center of the earth by the force of caloric exerting its self-repulsion throughout the universe.

Capillary Attraction—Is the cutting off of the converging rays of caloric from above by a small tube, leaving the rays from below free to carry the liquid upward.

Planetary Motion—Is the effect produced on the planets by the concentrating streams of caloric from the outside, acting on the rectilinear motion produced by explosion.

Electricity—Is caloric confined in channels or bubbles, or just escaping from them.

Positive Electricity—Is caloric passing in currents out of a body.

Negative Electricity—Is the same element passing in currents into a body.

Magnetism—Is the condition of a body around which an electric current is passing.

Polarity in Magnetism—Is the direction of the currents.

Transparency—Is such an arrangement of the particles of a substance that the rays of caloric pass through in straight lines, without carrying any gross matter with them.

Translucency—Is such an arrangement of the particles that caloric passes through pure, but not in straight channels.

Opacity—Is such a condition of a substance that caloric in passing carries with it matter too coarse to act on the sense of sight.

Cohesion—Is the pressing together of the particles of a substance by the self-repulsion of caloric from the outside.

Hardness—Is the close fitting of the particles.

Softness—Is the loose fitting of the particles.

Adhesion—Is the pressing together of two separate substances by the outside caloric.

Elasticity—Is the tendency of globules of caloric to assume a spherical shape, thus forcing the elastic substance into the form best fitted for that purpose.

Chemical Affinity—Is the breaking down of the cellular walls between different substances by the self-repulsion of caloric.

Combustion—Is the escape of caloric from ignited bodies.

Explosion—Is the sudden escape of a large amount of the same substance.

The Tidal Wave—Is the pressing out of the elastic sea, at both ends of the line of least pressure.

The Building Power of Animals and Plants—Is the self-repulsion of caloric, driving the inert matter of which the tissues are formed into the place marked out by the organization of the seed or embryo.

The Circulation of the Blood—Is the result of the caloric contained in it driving it through the veins.

Blistering, (either by freezing or burning,)—Is the breaking down of the cellular tissues by caloric going rapidly into or out of the flesh.

Contraction of the Muscles in Length—Is the result of their lateral expansion by the caloric admitted into them.

THE ONE GREAT FORCE of the Material Universe is the Self-repulsion of Caloric, acting on the Inertia of Ordinary Matter.

www.ingramcontent.com/pod-product-compliance
Lightning Source LLC
Chambersburg PA
CBHW021950190326
41519CB00009B/1206